TABLE OF CONTENTS

Radio waves

When current flows through a wire, a combined electric and magnetic energy "field" is generated, a zone of energy which extends at the speed of light into space. With alternating current (AC), the energy reverses forward and back 60 times per second (60 Hz). Each current reversal causes the magnetic (north and south) poles to reverse as well. This simulates a low-frequency radio wave.

The electric field ("E" for electro-motive force, measured in volts) is parallel to the axis of the wire, while the magnetic field ("H" named after researcher Joseph Henry) is perpendicular to it. This field is described as electromagnetic. Familiar illustrations depicting radio waves as wavy lines or crosshatched arrows are graphic representations only. There are no "lines of force" as implied when iron filings line up during magnet demonstrations; those filings line up because they all become little magnets, attracting and repelling one another.

Radio waves are only a continuous field of energy which, like a beam of light, is strongest at its source, weakening with distance as it spreads its energy over an ever-widening area. In fact, radio waves and light waves differ only in frequency over a continuous electromagnetic spectrum, with higher-frequency light having greater energy and the ability to be seen by some living organisms. Scientists even refer to an antenna as being "illuminated" by radio energy.

Radio waves can be reflected by buildings, trees, vehicles, moisture, metal surfaces and wires, and the electrically-charged ionosphere. They can be refracted (bent) by boundaries between air masses; they can also be diffracted (scattered) by a ground clutter of reflective surfaces.

Radio and light waves travel through the vacuum of space approximately 186,000 miles (300 million meters) per second, but when they pass through a dense medium, they slow down; this velocity factor, is given as a specification for transmission lines.

When we specify antenna and transmission line lengths, these are electrical wavelengths which are shorter than free-space wavelengths because of this reduction in speed.

Propagation

We refer to the behavior of radio waves as they travel over distance as propagation. Ground waves stay close to the earth's surface, never leaving the lower atmosphere. They are severely attenuated (reduced), rarely reaching more than a few hundred miles even under ideal conditions.

Surface waves, the lowest ground waves, often reach their destination by following the curvature of the earth. Space waves are the line-of-sight ground waves which travel directly from antenna to antenna. VHF/UHF space waves, when encountering abrupt weather boundary changes, experience temperature inversions and ducting as well as other influences; these can funnel signals into significantly extended ground-wave coverage.

At the upper reaches of our atmosphere, ultraviolet rays (UV) from the sun ionize (electrically charge) the air atoms, lending the name "ionosphere" to this highest zone of the earth's atmosphere. Radio waves which reach these ionized layers, 25-200 miles high, are called sky waves, The lowest regions of the ionosphere, the D and E layers, are influenced directly by sunlight; their effects begin at sunrise, peak at noon, and disappear after sunset. They absorb radio signals. The longer the wavelength (that is, the lower the frequency), the more the absorption. This explains why daytime reception below roughly 10 megahertz (MHz) is so poor.

But the E layer also reflects shorter wavelength (higher frequency) signals back to Earth; the higher the frequency, the more the reflection. That is what provides distance (DX) on the higher shortwave frequencies.

Most DX, however, is produced by the next region up, the F layer, which retains its electrical charge well into the night, reflecting signals back to the earth over great distances. All of these solar influences gradually increase toward a peak during the 11 year sunspot cycle, and then slowly diminish again.

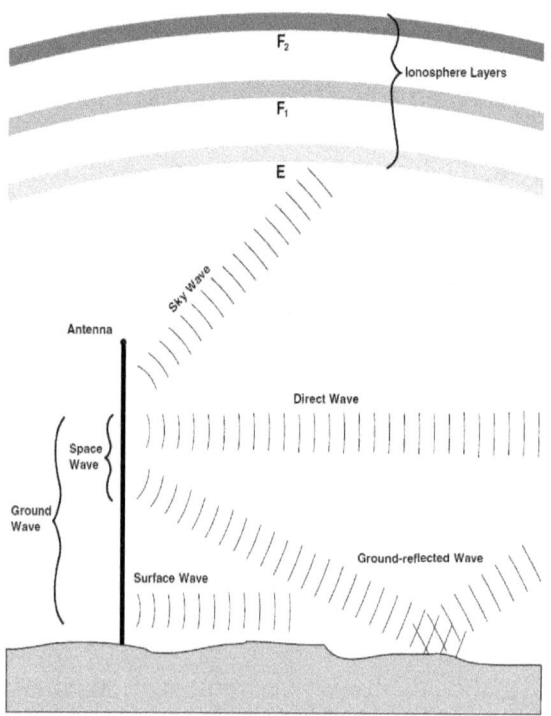

The earth itself can reflect radio waves, allowing a phenomenon called multihop ("skip"); combinations of earth reflections and ionospheric refractions producing as many as five skips. More bounces than that would be attenuated by ionospheric absorption and terrestrial scattering, rendering the signal unreceivable.

Internet sites publish continuously updated radio-propagation forecasts and a variety of prediction computer programs, allowing the user to plan ahead for the most productive use of the spectrum.

It is estimated that some 200 tons of meteor material, from visible to dust size, strikes the earth every day; much more vaporizes in the upper atmosphere. Because of this constant bombardment, completely automated systems rely on meteor scatter for long-distance data transfer.

Terrain, trees, wiring, metal siding, nearby buildings, and other reflective surfaces all affect antenna performance. The lower the antenna, the more obstructed it is likely to be. A basement would be a very poor antenna location. Signals are unpredictably reflected by metal and wiring in and on the walls and ceiling; nearby electric and electronic appliances invite interference to reception; soil absorbs transmitted energy and also reflects signals upward; and signals come mostly from overhead rather than from the horizon.

Patterns

The shape of the field of energy emitted by a transmitting antenna, as well as the geometric response by a receiving antenna, is known as its pattern. It may be a simple donut shape surrounding the axis of the wire as in a half-wave, or smaller, dipole, called a doublet, or it may be multi-lobed, as in a multiple-wavelength antenna, called a longwire.

Nearby trees, buildings and hills take their toll, too. Locating an antenna inside a large building with steel frame and metal reinforcements may attenuate signals up to 25 dB at VHF and UHF, according to one study. Brick walls, slate or tile roofs can account for 6 dB, even more when wet. Shorter wavelengths (900 MHz) get through small windows in shielded walls where longer wavelengths (150 MHz) do not.

It would be nice if we could simply assume that pointing a directional HF antenna toward the horizon would result in a zero radiation angle with respect to the Earth's surface. But phase relationships from ground reflections elevate the takeoff angle.

The resulting takeoff angle is a mix of near-field considerations like frequency (wavelength), initial angle of the radiation from the antenna element(s), and height above ground.

In the far field, waves may reflect and recombine in or out of phase, thus enhancing or diminishing the signal strength in some planes. Propagational effects of the atmosphere and the ionosphere absorb, reflect, and refract the waves.

Antennas are designed to favor certain directions, both for transmitting and receiving. The lower the frequency, the more the signal is capable of following the contour of the terrain, and the less likely it is to be absorbed by trees and foliage. One study showed that with dense trees and vertical polarization, attenuation at 30 MHz is about 3 dB, increasing to 10 dB at 100 MHz.

The elevation pattern, variously called radiation angle, takeoff angle, maximum amplitude elevation, and launch angle, is affected by height above ground, length of the antenna element(s), and the presence of nearby metal, including other antenna elements. It is an integral part of an antenna's gain characteristics.

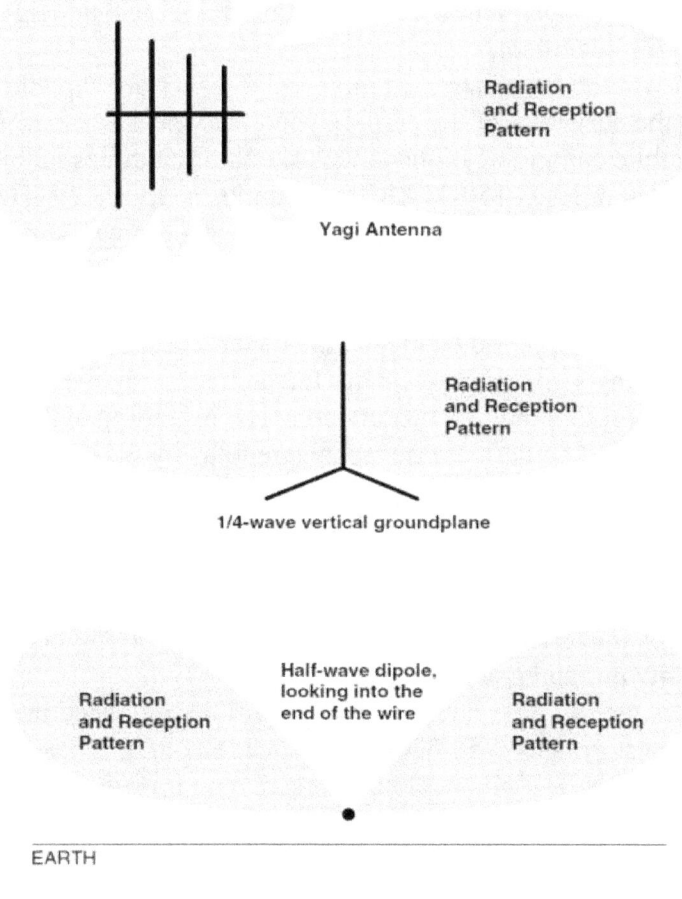

Radiation and Reception Pattern

Yagi Antenna

Radiation and Reception Pattern

1/4-wave vertical groundplane

Radiation and Reception Pattern

Half-wave dipole, looking into the end of the wire

Radiation and Reception Pattern

EARTH

The Radio Horizon

Radio waves, like light waves, follow the line of sight. Because of the curvature of the earth, higher antennas "see" a farther horizon. Assuming a flat, unobstructed terrain, the visual horizon is about 8 miles for a 30-foot-elevated antenna, increasing to only 16 miles at 120 feet. Notice the square law effect – it requires roughly four times the height to get twice the distance. Once an antenna is high enough to "see" past nearby obstructions, it takes at least double that height to notice any improvement.

The lower the frequency, the more radio waves are capable of following the curvature of the earth beyond the visual horizon. Typical base-to-mobile communications ranges are about 50 miles in the 30-50 MHz band, 30 miles at 150-174 MHz, 25 miles at 450-512 MHz, and 20 miles at 806-960 MHz. Obviously, these distances will vary depending upon radiated power, receiver sensitivity, antenna gain, elevation and location.

Although the higher the antenna the better, coax cable losses may compromise any signal improvement; the higher the frequency, the worse those losses. For example, at 450 MHz, extending a 30-foot antenna to 60 feet could increase signal strengths by 5 dB, but if you are using common RG-58/U coax, signal strengths may be attenuated by the same amount, resulting in no improvement at all.

At 800 MHz, using this small diameter, lossy RG-58U, signals would get worse with height. Worst of all is thin RG-174U which has all the bad characteristics; in long lengths at UHF, you might as well short-circuit your antenna connector.

Always use low-loss cable such as the following, listed in increasing performance: (50 ohm impedance) RG-8/X, RG-8/U, Belden 9913, or Andrews 1/2" foam, Heliax; or (72 ohm impedance) RG-59/U, RG-6/U, or RG-11/U).

Low or High Frequencies?

Radio waves have a tendency to propagate differently for different frequency ranges. At very high and ultra high frequencies (VHF and UHF), they behave more like light, travelling in a straight line. At the lowest frequencies, they tend to follow the curve of the earth, through it as well as over it.

For listening to very low frequencies (VLF) below the AM broadcast band, a simple vertical antenna may be ground mounted for reception of beacons. In fact, a long, random wire can be simply draped over the ground, or even buried an inch or two under the soil for satisfactory monitoring.

If directivity is required, an open-wire loop or a ferrite rod antenna may be used. Such antennas are effective up through the AM broadcast band and even into the first few megahertz of shortwave.

Most shortwave listeners choose horizontal wire antennas. They are the least expensive, they require simple suspension systems and they can be oriented to favor specific directions. But they do require a fair amount of yard space and can be conspicuous in restrictive residential neighborhoods.

Vertical antennas on these bands are omnidirectional, thus receiving signals equally from all compass points. This is a good thing when sheer signal abundance is the object, but can also compound co-channel interference among users of the spectrum who happen to be transmitting on the same frequency.

At shortwave (more correctly, high frequency or HF), distant (DX) signal patterns have been altered substantially by the time they arrive, and vertical or horizontal antennas are chosen more on a basis of convenience. Each has its retinue of loyalists who claim that one is better than the other.

What is a ground?

The earth plays an important role in radio signal propagation, but *grounding* your radio equipment is not one of them. Attaching the chassis of your radio to a buried conductor in moist soil may protect you from electrical shock, drain off static-charge buildup, help dissipate nearby lightning-induced spikes, and even reduce electrical noise

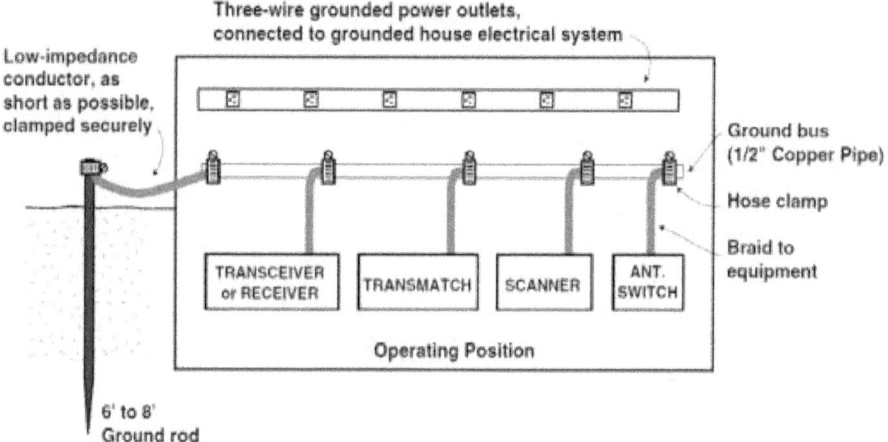

A good ground system utilizes short, large-gauge wire to connect radio equipment commonly to at least one deep ground rod.

A good electrical ground consists minimally of two eight-foot metal rods, at least ten feet apart, connected to the radio equipment by a short length of heavy braid. Moist, mineralized soil is best; dry, sandy soil is worst.

It's a good idea for all of your protective wiring to have only one attachment point to your ground. This avoids potential differences arising at different attachment points.

The RF ground

A radio-frequency (RF) ground, on the other hand, is more extensive. A vertical antenna may be thought of as a center-fed dipole turned on its end, and the lower half removed so that we can mount the remaining element on the ground where the coax will be attached. But we must somehow supply that missing half of the antenna.

If we merely bury the needed wire or attach a ground rod (B), the energy that would radiate from that element is absorbed by the mineralized soil, simply heating it. Such an element is sometimes referred to as a "worm warmer."

Instead, we construct a *counterpoise* on or above the soil (A), a metallic surface emulating a "perfect" (reflective) earth, composed of radial wires connected to, and extending outward from the coax shield at the base of the antennas.

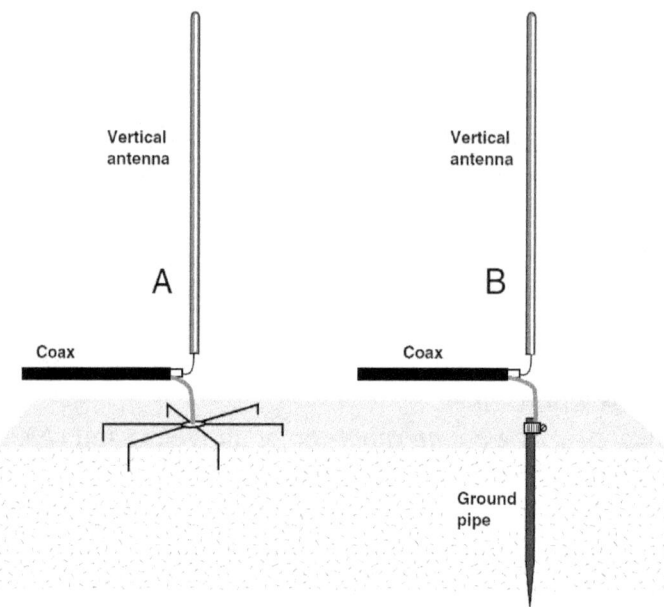

How many spokes of wire, and how long? AM broadcast stations use at least 120 radials for transmitting purposes; you should use at least 16 1/8-wavelength wires to avoid power losses from soil absorption.

Because current is at its maximum at the feed point, density of metal around the base of the antenna is more important than the length of the radials. If you have 100 feet of wire, ten 10-foot lengths are better than two 50-foot lengths.

This is not so critical on receive-only antennas. Even a single quarter-wavelength wire provides counterpoise effect; it may be run randomly or even coiled loosely in some cases. Such a wire is often connected to the chassis of the transmitter if it is "hot" during transmitting as evidenced by painful RF burns when touching the equipment, especially your lip to the mike!

Construction

Two neighboring shortwave listeners decide to erect antennas to monitor 41-meter (7.1-7.3 MHz) international broadcasting. One neighbor, using rocks as counterweights, throws about 50 feet of small-gauge hookup wire over a couple of tree limbs; it sags in a number of places, has no insulators other than its plastic covering, and averages some 30 feet in the air. At the center cut of the wire he has soldered a 50-foot length of TV coax which he runs down to his receiver.

His neighbor, a purist, erects two 30 foot telephone poles 60 feet apart, stretching 66 feet of heavy gauge, silver plated, uninsulated wire between porcelain insulators. The antenna is in an open yard with no trees. At the center, he carefully attaches a commercial coax connector, from which he runs a 50-foot length of large-diameter, low-loss, RG-8/U coax.

Does the purist hear signals any better? Nope. Assuming identical environment and antenna orientation, reception will be virtually the same. The difference in signal strengths between 50 and 66 feet is imperceptible, even though the 66 feet represents a closer impedance match at 7 MHz.

The plastic-coated wire insulates it from the moist tree limbs, but even if it touched, the resistance of the trees would not contribute significant signal loss. Signal absorption by foliage at 7 MHz is minimal; the resistance of the thinner wire is less than one ohm; and the difference between 50 feet of RG-58/U and RG-8/U at 7 MHz is a mere fraction of a dB.

For receiving purposes, an antenna may be thick or thin; its texture may be solid, stranded or tubular; its composition may be any conductive metal (gold, steel, copper, lead, or aluminum); it may be covered with insulation or left bare. All signals will sound virtually the same.

Even if signal strengths were reduced considerably, they would still be just as audible, because at shortwave frequencies, once there is enough signal to be heard above the atmospheric noise (static), a larger antenna will only capture more signal *and noise*. The S-meter may read higher, but you would hear the same signal above the noise audio with the "deficient" antenna by simply turning up the volume control.

So why bother with good construction practices? Heavy gauge, stranded wire will withstand ice, wind loading, and flexing better than thin solid wire, and it will radiate transmitted power more efficiently.

Commercially made center insulators with built-in connectors are more rigid and water resistant than soldered connections and they can be easily disconnected for servicing or inspection. Sturdy, insulated suspension is more durable over time, and keeping antennas away from tree foliage may avoid some signal loss at higher frequencies.

Skin Effect

A thin, hollow, metal tube is just as efficient in conducting and radiating radio-frequency energy as a solid wire of the same diameter and material. This is because RF energy barely dips below the surface of the conductor, and the higher the frequency, the shallower the depth.

The larger the surface, the less resistance, which would waste power as heat. Skin depth varies inversely with the square root of the conductivity and the permeability (magnetic attraction) of the metal; the better the conductor, the deeper the skin effect.

At microwave frequencies (10 GHz), the skin depth of silver, an excellent conductor, is 0.64 micrometers (μm), while that of aluminum, a poorer conductor, is 0.80 μm.

Iron is a very poor conductor and has high permeability; its skin depth is only 1/7 that of copper, and it rusts in the open, making it a poor choice as a conductor at radio frequencies.

Antenna Size

The energy-intercepting area of an antenna is called its *aperture* (capture area), another similarity to light as in the aperture of a camera lens. The larger its aperture, the more signal it captures.

Curiously, a large antenna is not necessarily better at transmitting (or receiving) than a smaller antenna. If a small element can be designed to be just as efficient as a large antenna, and radiates the same pattern, there is no benefit in using a larger antenna unless it can be configured to offer *gain*, which comes from shaping the directionality of the antenna.

Similarly, all antennas of the same size (wire dipoles, folded dipoles, fans, trap antennas, cages, or any others) radiate the same amount of power. Their relative advantages come from pattern directivity.

A common question, often brought up during antenna discussions, is whether a good transmitting antenna is also a good receiving antenna.

The answer is yes if it's long enough for incoming signals to overcome the receiver's self-generated noise level. Once that has been accomplished, any further signal magnification brings with it natural atmospheric noise and electrical interference which accompanies all signals in this part of the spectrum.

The U.S. Coast Guard found in the 1950s that a five-foot antenna was adequate for HF reception 100% of the time. Then the corollary arises: Is a good receiving antenna also a good transmitting antenna? That answer is, not necessarily.

Transmitting antennas are designed to radiate as much of the transmitter-generated signal as possible. This means that special consideration must be given to the wire gauge, impedance matching, line loss, and any other power-robbing elements of the system.

Wire antennas for the 2-30 MHz shortwave range that perform as well as expensive commercial antennas are easily constructed using simple, inexpensive materials.

Placement

Whether you're transmitting or receiving, the positioning of any antenna is of vital importance. Unless the earth is being used as part of the grounding element of the antenna as it often is with verticals, it should be high and in the clear. It would be hard to think of a worse location than in the basement of a metal-sided building!

A vertically polarized antenna can be mounted at ground level, away from obstructions, on a pole or tower, or even on a rooftop. Most HF verticals require a ground counterpoise, an array of conductors radiating from the base to provide the reflection that an earth ground is incapable of producing. Dirt is not an efficient conductor, it is more like a giant resistor, especially dry sand or clay soils!

Horizontal antennas, on the other hand, are balanced systems requiring no ground. They do, however, have to be suspended high enough above the ground to avoid whatever reflection that soil has.

A horizontal antenna close to ground reflects upward, making it fine for communicating with an aircraft overhead, but not for much on the horizon!

Filters

There are occasions when you wish you could reduce the strength of a local broadcaster, weather channel, CBer, paging transmitter, or other source of strong signal overload. Frequency-selective notch filters and band pass filters are available for these instances, and some manufacturers will even provide custom frequency units.

Preamplifiers

We've listed the caveats against boosting shortwave signals since some of them are already very strong, and receiver front-end overload produces many negative characteristics such as desensitization, intermodulation, and images, all of which tend to reduce rather than enhance reception. But there are fixes.

A preamplifier ("pre-amp" or "signal booster") is simply a small-signal amplifier placed between the antenna and receiver. When integrated with a small receiving antenna, the combination is called an active antenna.

Preamps connected to poorly-located antennas will not perform as well as well-placed, larger "passive" (unamplified) antennas, but they may be the only alternative when better antennas are not practical.

Ideally, a preamp should be mounted at the top of the mast, right at the antenna's feed point. This way you know you are going to boost the signal substantially, probably overcoming absorption from the loss in the coax.

If you put the preamp all the way down the transmission line just before the receiver, weak signals may be lost in the coax and are no longer available to boost. All you'll get in that case is louder "hiss."

Keep in mind, however, that these preamps cannot withstand a transmitter's power. They must be bypassed or disconnected from the antenna line during transmission.

If a shortwave receiving antenna is at least 20 feet long and in the clear, a preamplifier is probably unnecessary.

A preamp must have a lower noise figure (self-generated "hiss") than the receiver, or the only thing it accomplishes is increasing both signal and noise, just as if you had merely turned up the receiver's volume control.

It must have wide dynamic range – the ability to amplify weak and strong signals equally without becoming overloaded and thus generating spurious signal products, known as intermodulation (intermod), which interferes with normal reception.

Intermod and images are different from phantom signals ("birdies") which are dead carriers (unmodulated signals), products of the receiver's own oscillator circuitry, and are heard in the wrong places on the tuning dial. They are present whether or not a preamplifier is used.

One control for overload is a variable attenuator, essentially an adjustable resistor, placed between the output of the preamplifier and the receiver's antenna jack. It can be adjusted to reduce the amplified signal low enough to avoid overloading the receiver, while still boosting signal strength.

Such attenuators are available at low cost from department stores that sell TV accessories.

At VHF and especially UHF frequencies and above, where transmission line losses may become significant, a preamplifier mounted at the antenna will boost signals above the loss characteristic of the line. Still, the preamp is vulnerable to all the problems described above. Use it as a last resort.

Preselectors

But preamplifiers with preselection are another matter. Most are equipped with gain controls to avoid over-amplification, and the sharp tuning of the preselector homes in on one very narrow part of spectrum in order to boost only those weak signals. Of course if you already have a tuner (passive preselector), then you can simply add a wideband preamp and choose the small signal swaths that you desire to amplify.

A receiver preselector

Noise Sources

In our modern electronic age it's no wonder that electrical and electronic interference surrounds us. Power lines arcing, computers, microprocessor-controlled appliances, Broadband Over Power lines (BPL), switched-capacitance power supplies, plasma TV sets and other emitters of interference can pollute the radio spectrum you're trying to hear.

In the early days of mobile radio, only spark ignition noise and alternator whine gave us a problem, but now with computerized ignition systems, there is a spectrum of hurt awaiting mobile radio enthusiasts.

In reality, the majority of these noises dissipate the higher in frequency we tune, but the bad news is that much of it is still present well into the shortwave spectrum. And if you happen to live in an apartment complex, my condolences!

Noise reduction

But not all is lost just because you are surrounded by electrical noise generators. There are measures that can be taken to reduce such annoyances. Ideally, of course, your antenna should be outdoors and away from buildings and power lines. That's the first measure for noise interference reduction.

But modern receivers have been designed to cope. Automatic noise limiters are the oldest countermeasure for electrical noise, and they still work. Noise blankers work by momentarily turning off the receiver when a noise pulse is detected; it happens so fast that you hardly notice it.

Digital signal processing (DSP) is more sophisticated, converting the received analog signal into digital bits so that the noise portion can be selectively removed.

Another interesting device is an accessory product descriptively known as a noise canceller. It works by using two antennas. One, the signal antenna, and the other, called a noise antenna.

Since the noise arrives at different times at the two antennas, their relative phase angles may be compared and subtracted from an adjacent portion of the spectrum while not affecting the frequency to which you are listening.

Lightning protection

Warm-weather storms bring hazards to radio enthusiasts that most of the public doesn't even think about. Tall metal objects (antennas and their masts and towers) present an open invitation to lightning strikes. Hundreds of millions of volts at hundreds of thousands of amps – that's a wallop!

Many of us tend to think that since it hasn't happened to us before it's not likely to in the future. The same thoughts often occur to those who haven't had a previous car accident.

Ever since Ben Franklin invented the lightning rod, bolts from the blue have been hitting those conductors and (hopefully) travelling safely to the ground.

So what happens when lightning hits an antenna that is connected to expensive radio equipment? We needn't discuss that; let's just talk about lightning protection.

Broadcasting stations have elaborate lightning protection, including extensive grounding of supporting towers, guy wires, and coaxial cable shields.

Some readers right now are recalling a lightning-strike episode. Even a nearby strike can cause serious damage through induction between the bolt and the conductive antenna system; their close spacing is like a giant air-core transformer.

The surest form of equipment protection from nearby lightning strikes is to disconnect all antennas from the radios. Early hams used to remove the connector from the rig and dangle it in an empty drinking glass for additional insulation from surrounding objects. But some storms sneak up on us, or we may be distant from our homes when an unannounced weather system moves in.

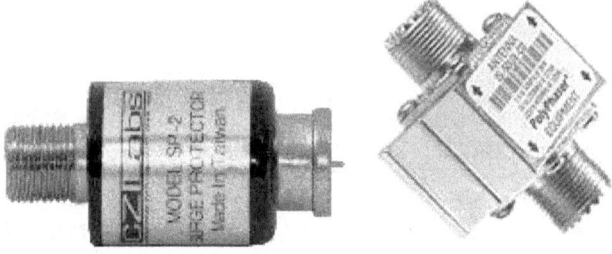

Lightning arrestors are made for a variety of applications; I'd recommend the gas-discharge type which can be inserted right in the coax line and connected to a solid ground wire. Some models of antenna switches include a grounding position to connect the antenna line directly to ground when not in use. Good idea.

Due to the speed of the lightning discharge, it has high-frequency characteristics. Wrapping a dozen or so turns of the coax into a coil before it enters the dwelling isn't a bad idea, and running that coax through about ten feet of grounded metal pipe is another. Both of these measures can be thought of as RF chokes.

But what if the lightning hits house wiring? Your equipment may be disconnected from the antenna, but it's still plugged into the wall. Be sure your home electrical system is protected at the breaker box by a husky metal-oxide varistor (MOV) transient voltage suppressor.

Even your telephone wiring is vulnerable. Get yourself one of those surge protection outlets that has a modular telephone protection jack as well.

Gain

The concept of gain may seem rather elusive, especially when you read the claims of manufacturers. The unit is the decibel (dB), the improvement in signal strength over a standard antenna.

That standard may be one of two: an imaginary, ball-shaped (isotropic) antenna which would radiate uniformly in all directions, but doesn't exist in the real world; or a half-wave dipole, often configured as a quarter-wave antenna mounted on a quarter-wave counterpoise (ground plane) of radial elements.

To let the reader know which standard the product is being referenced to, the gain must be stated in the appropriate unit, dBi (isotropic) or dBd (dipole).

The dipole radiates from its sides, not its ends, providing 2.1 decibels of gain (2.1 dBi) over the isotropic model (0 dBi). Manufacturers prefer to specify their antenna gains in dBi since it will be 2.1 dBi higher even though it's a fictitious reference. Beware of products that give only the number and not the unit.

Increase in gain comes from shaping the radiation/reception pattern, but adding to the pattern in one direction means reducing it in another. Such pattern re-direction often refers to front-to-back ratio and side-lobe rejection.

The pattern can be shaped by adding parasitic elements, which are unconnected but secured to the boom, called reflectors and directors. This is the structure for a beam antenna, more correctly called a Yagi.

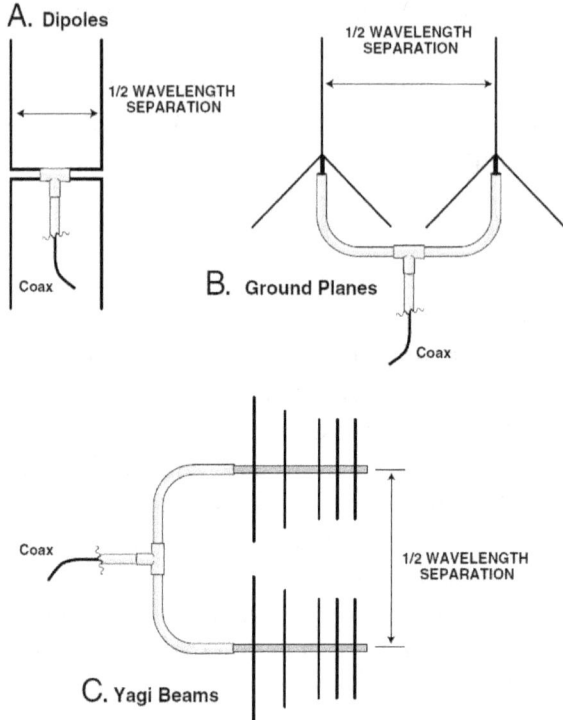

A. Dipoles

1/2 WAVELENGTH
SEPARATION

Coax

1/2 WAVELENGTH
SEPARATION

B. Ground Planes

Coax

Coax

1/2 WAVELENGTH
SEPARATION

C. Yagi Beams

Because this can get cumbersome at low frequencies with their longer wavelengths, beams are generally reserved for the VHF/UHF frequencies and the high end of the shortwave spectrum, typically above 14 MHz.

Adding a second identical antenna separated by ½ wavelength and connected in phase, known as stacking, will increase transmitted and received signal strengths by 3 dB, regardless of the original gain. Thus, two 1-dB-gain antennas will provide 4 dB total gain (a 3 dB increase), and two 20-dB-gain interconnected antennas will provide 23 dB total gain (still a 3 dB increase).

In a mobile whip, this can be done by co-phasing an upper and lower element on a continuous whip. This is known as a collinear (two elements in line) antenna. The sections are isolated from one another by a coil. Such antennas can deliver roughly two decibels of gain, barely noticeable except on extreme fringe.

Configuration

Whether you're considering a single-element antenna, multiple antennas for frequency agility, or a more complex array for gain and directivity, the choices are abundant. The most common shortwave antenna is the center-fed dipole (flat top, doublet, T-aerial), a horizontal wire fed at the center. It has a pattern favoring signals coming in (and going out) at its sides, while nulling (minimizing) the pattern at its ends.

If you leave one end of a dipole high, and lower the other to just about the ground, you have a "sloper" which favors signals, both transmitted and received, in the direction of the downward slope.

The inverted-V and inverted-L are descriptive of two wire antenna configurations resembling those two letters seen upside down. Their singular advantage is to save space without a consequential reduction in performance.

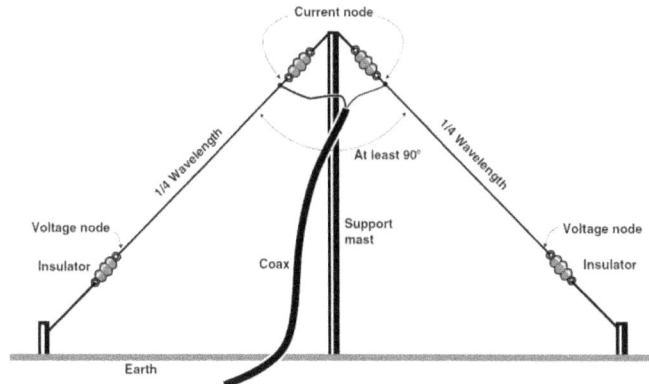

Longwires are long in terms of wavelength. A 136 foot horizontal wire may be a half wavelength on 80 meters, but it's a longwire by two wavelengths on 20 meters.

A random wire antenna is just what its name suggests. It is not planned for a specific frequency, but simply long enough to get good reception and be able radiate well when impedance-adjusted with a transmatch.

Wire

Whether you are planning to build a rotatable beam or simply stretch a wire, a bill of materials must be anticipated.

Wire antennas, most often variants of a center-fed, half-wave dipole at the frequency of interest, may be solid or stranded. To protect against wind, falling branches, and ice, use stranded wire for its flexibility. It makes absolutely no difference whether the wire is insulated or uninsulated, the plastic covering is transparent to radio waves.

Some bare wire comes with enamel coating making it less susceptible to corrosion. Of course the enamel must be scraped off for soldering at the feed point. After the connections are made, the bare, soldered joints should be sprayed with acrylic to resist corrosion.

A wire gauge of #14 is quite satisfactory for nearly any antenna design. Since it will come as a spool or coil, don't stretch it out like a giant spiral; unroll it a turn at a time so that it lies flat. This prevents it from kinking, thus weakening it under strain of wind and ice load.

Copper-clad steel wire is the toughest, but a little more difficult to manage since it is prone to spiraling unless pulled tight. Even aluminum clothesline, which can't be soldered, will work provided the mechanical joints between the aluminum wire and copper feedline are well coated to prevent electrolytic corrosion.

Insulators

Glass and porcelain insulators are used at the ends of wire antennas for suspension. The wire element should be sized to length from the feedline connection to the end loop where it bends through the insulator hole and back on itself to be twist-wrapped on the wire. It's a good idea to solder that twist wrap for security, then spray the soldered twist with acrylic to resist corrosion.

On the far end of the insulator, another hole is provided for a tether. It may be wire, but polypropylene and nylon are easier to work with. The tethers may be wrapped around secure tree limbs or trunks, or even led over a limb, then down to the base of the trunk for easy raising and lowering of the antenna.

So how do you get the tether up that high in the first place? Many seasoned veterans simply tie a rock to the end of a thin rope and throw the rock over a limb. Others have even used a bow and arrow on a string. EZ Hang even makes a slingshot to get that line over a limb. Then there's the rare individual who simply likes to climb trees!

In the absence of trees, commercial sources of poles, masts and towers are available.

Transmission Line

During the 1950s, TV twin-lead and its superior cousin, ladder line (also known as "window line"), were commonly used as transmission lines for radio communications. These flat "ribbons" were more tolerant of impedance mismatch , but still had to be run in the open to avoid bending and coming close to conductive surfaces.

Even so, they worked with vacuum-tube transmitters which could stand the high voltages produced by voltage reflections from poorly-impedance-matched antenna systems. At about the same time, coaxial cable, which came into military prominence during World War II, found its way into amateur radio and other domestic communications.

While flat line (window line, ladder line, twin lead, open wire) is traditionally granted the lowest loss characteristics, it is vulnerable to picking up radio frequency interference (RFI) from nearby power lines and electrical/electronic appliances.

Like unshielded, single-wire feed lines of the early days, flat line can't be run against metal surfaces, or across wiring, or in moist ground, or bent sharply, it's difficult to navigate through walls to the radio shack, it changes impedance when it gets wet, it gets lossy as the weather dries it out and cracks it, and there are no inter-series adapters. And once it's there, it requires a balun (balanced to unbalanced) transformer to mate with the unbalanced coax connector (typically an SO-239) on the radio equipment.

Coaxial cable is far friendlier for transmission line. The most commonly used coax for typical 50 ohm systems are RG-8/U and RG-58/U, both of which attach to conventional PL-259 (sometimes called "UHF," a misnomer) male connectors. RG-213/U is a more ruggedized, mil-spec version of the RG8/U, and is overkill for most ham radio applications.

RG-58/U is thinner and therefore far more flexible than RG-8/U, and is well suited to power levels in the 100 watt range of typical HF transceivers. As with any transmission line, the higher the frequency, the higher the loss in the form of heat across the dielectric (inner insulation). But even on 10 meters (28 MHz band), the loss for a 100 foot length is only about 2 dB, barely noticeable even on the weakest signal level.

 RG-8U

RG-8/U in comparison loses about 1 dB to heat at that frequency, and it's virtually impossible to hear that loss which is also only 1 dB less than RG-58/U. However, for longer line lengths and higher power with an external linear amplifier, the RG-8/U justifies its cost.

So what about RG-6/U, the low loss, low cost cable used in the TV industry? It handles 100 watts, has about the same loss characteristic as expensive RG-8/U, and even though it best fits the type F screw-on TV connector, low-cost adapters are readily available for any other mating duty including PL-259/S0-238 requirements on all HF transceivers.

It also has 100% shielding, making it more immune to extraneous RFI pickup than RG-58/U which does not have 100% shielding. Of course, purists will argue that there will be some additional loss because the TV coax is 75 ohm, not 50 ohm.

What they don't argue, though, is that over the wide frequency ranges that hams use their antennas, the system impedance doesn't remain 50 ohms, either. That's where "tuners" come in.

As a general rule, the transmission line should not run closely parallel to an antenna element; this distorts the pattern. For dipoles, a good rule of thumb is to run it out at a right angle at least a quarter-wavelength at the lowest critical frequency, and then it can be curved into any direction (except back toward the antenna) with minimal distortion of the pattern.

Although the higher the antenna the better, coax cable losses may compromise any signal improvement; the higher the frequency, the worse those losses. For example, at 450 MHz, extending a 30-foot antenna to 60 feet could increase signal strengths by 5 dB, but if you are using common RG-58/U coax, signal strengths may be attenuated by the same amount, resulting in no improvement at all.

At 800 MHz, using this small diameter, lossy RG-58U, signals would get worse with height. Worst of all is thin RG-174U which has all the bad characteristics; in long lengths at UHF, you might as well short-circuit your antenna connector.

Always use low-loss cable such as the following, listed in increasing performance: (50 ohm impedance) RG-8/X, RG-8/U, Belden 9913, or Andrews 1/2" foam, Heliax; or (72 ohm impedance) RG-59/U, RG-6/U, or RG-11/U).

Connectors

  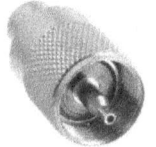 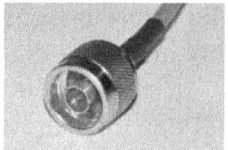

Popular F, BNC, PL-259, and N male connectors

Coax connectors come in a variety of styles, including PL-259, N, BNC, F, SMA, and TNC. But for the shortwave spectrum, the PL-259 plug and its mated SO-239 socket are universally preferred.

The end of the coax is first trimmed step by step, revealing the length of bare center wire, the inner braided shield, and the terminal edge of the outer vinyl jacket.

For RG-58/U cable, the PL-259 employs a reducing sleeve to accommodate the smaller cable diameter. The sleeve is not used with RG-8/U. But both require solder, and 60 percent tin, 40 percent lead is the most common alloy.

Electrical solders usually have a core of rosin which aids the process by preventing oxides from depositing on the connection while the solder is molten. Never use acid core solder on electronic connections; it will etch and corrode metal components.

Antenna Switches

It is often desirable to select among two or more antennas for optimum reception or transmission. For receiving purposes, or even for low power (a few watts) transmitting, most TV coax antenna switches work admirably from DC through 1000 MHz.

CB-type antenna switches work fine up to about 30MHz, for both receiving and transmitting. For higher power, especially at higher frequencies, select a commercial coax switch rated for the frequency range and power required.

Splitters and combiners

A splitter is essentially a broadband RF transformer which allows one signal source to be equally divided into two or more paths. This allows two or more receivers to operate from one antenna. Since a two-way splitter is a power divider, each output will be reduced by 3 dB, half the original power level. There will also be some minor additional loss from the resistance in the windings.

Connected in reverse, a splitter becomes a combiner, allowing two signal sources to add commonly. This allows, for example, two separate-frequency antennas to be used simultaneously with one wide-spectrum receiver.

But there is a caveat for doing this. If the two antennas have a similar frequency response, they can produce destructive interference (signal canceling) from signals arriving out of phase from certain directions.

This same anomaly can be an advantage as well – it provides 3 dB overall gain in other directions! Basically, the system comprises a directional array – a "beam" antenna. TV splitters marked "V/U" or "VHF/UHF" or "54-890 MHz" actually work reasonably well from the low HF range (typically around 3 MHz) up through 1 GHz or more.

For transmitting, TV splitters will also allow low power – a few watts – to pass without much problem, but higher power levels will heat the fine winding and saturate the small ferrite core, wasting power and even destroying the device. Transmitter splitters and combiners are also available, but due to their rugged requirements, they are substantially more expensive.

Tuners

For maximum performance of any antenna, the transceiver needs to match the impedance of the antenna system. This is commonly done with a transmatch, more popularly known as an antenna tuner.

But this reference is misleading. Nothing in the shack can tune an antenna.

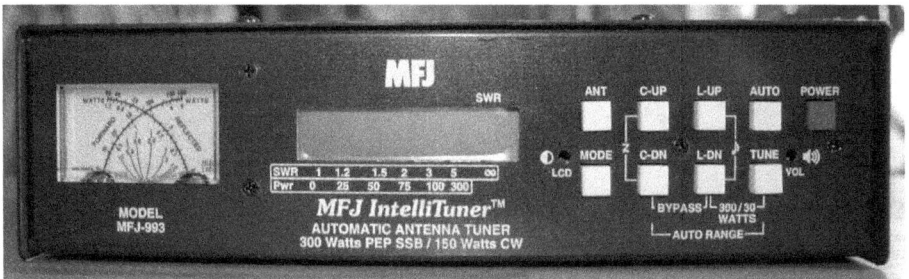

While the feed point impedance of any antenna will change over a range of frequencies, the feedline impedance never changes. What the transmatch does is to add capacitance and/or inductance to offset that mismatch at the rig. It's called conjugate matching. The impedance is still mismatched at the antenna, but it looks right at the transceiver.

Balun Transformers

What is a balanced antenna? One that is symmetrical on each side of its feedline, and electrical voltages and currents are identical, although in opposite phase, in each wire. A center-fed dipole and most beam antennas are excellent examples. They should be fed with flat line.

But we most commonly feed any antenna with coax, so what happens? The imbalance between the balanced feedline and the unbalanced antenna causes some current to flow on the outside of the coax shielding which is radiated, thus reducing the apparent gain of the antenna with wasted, randomly radiated, signal power.

For reception, the imbalance causes a loss in symmetry and thus predictability in the reception pattern.

The common cure is the balun (balanced-to-unbalanced) transformer. It can be a stand-alone accessory, or may be implemented in a transmatch. Most commercial baluns have a pair of wire terminals, screw eyes or screws on one end for attachment to the elements of a dipole, and an SO-238 connector on the other end to mate with a conventional PL-259 male connector on the coax.

Depending on their turns ratio, balun transformers can have impedance matches of 1:1 and much higher. For example, a 50 ohm cable would be attached to a 200 ohm dipole feed point with a 4:1 balun.

Baluns are very broadband by nature and can be easily made by winding a few turns of wire through a ferrite toroid. They can also be purchased as an easy kit, or bought fully assembled with connectors ready to install on the antenna.

Antenna Q&A.

Q. I'm only interested in listening to distant shortwave signals. What's the least expensive way to design an antenna for this application?

A. Any kind of wire – stranded or solid, thick or thin, insulated or uninsulated, aluminum clothesline, steel or copper – will work. Yes, I can hear it now: "Aluminum corrodes, steel rusts, and thin wire breaks." That's true, but all of them receive the same.

But for best results, use 30-50 feet of convenient, reasonably strong, copper wire. Locate it away from your house, away from power lines, and elevate it as high as practical, positioned with its sides facing the target direction.

Solder it at the near end to any type of coax to your receiver. Be sure the shield of the coax is connected to the receiver ground (connector shield or chassis) as well as the center conductor.

Q: **I live in an underground apartment with metal siding and roof. I'm not allowed to put up an outside or attic antenna. What can I do to get good reception?**
A: Move.

Q. Are antennas designed for transmitting also good receiving antennas?

A. Except for the smallest loops, absolutely. But the converse isn't true; just because an antenna receives well doesn't mean it is the best choice for transmitting. Transmitting antennas are meticulously adjusted for the best impedance match with the transmitter, but just because you hear signals well on a randomly-designed antenna doesn't mean it will properly match a transmitter.

Q: Can a random receiving antenna be used for transmitting as well?

A: Yes if you used a "tuner" (transmatch) to match the nominal 50 ohm impedance of the transmitter to that of the antenna system. A transmatch adds inductance and/or capacitance so that the transmitter "sees" 50 ohms impedance (conjugate matching).

To maximize efficiency and effectiveness, use thick wire, low-loss coax, and a length of antenna element that is easily matched by a transmatch.

Q. I have a chance to buy a used antenna that's presently on a tower for only a fraction of its original cost; is that a good investment?

A. Only if it isn't corroded or weather-beaten. Check it thoroughly for signs of abuse, damage, repair, rust, and other corrosion. Generally speaking, antennas near saltwater beaches corrode faster than those farther inland. If it passes the test, have at it! But, safety first: To remove an antenna from a tower may require you to rent a "cherry picker" or hire experienced tower climbers with appropriate safety gear to do a safe removal.

Q: I know that shortwave antennas are pretty easy to design with only a 2-30 MHz frequency range, but what about VHF/UHF scanners with a 30-1000 MHz typical span?

A: There are four ways commonly used to widen VHF/UHF coverage on a single antenna: discone design, log periodic beams, trap antennas with coils to electrically separate resonant lengths, and multiple element lengths joined together at the base. All these techniques are useful for transmitting as well as receiving.

Q. I want to hear stations from the greatest distance over the widest frequency range on my scanner; what kind of antenna do I want?

A. For that demand, you'll need a log-periodic dipole array (LPDA) and a rotator on a tower, connected to your scanner with low-loss cable.

Q. For VHF/UHF scanner listening, do I want a vertical antenna or a horizontal antenna?

A. Vertical to match transmitting antennas at base locations on mobile units. Even a handy-talkie works more reliably when held in a vertical position, matching the units with which it is communicating.

Q. How about shortwave listening and HF transmitting? Vertical or horizontal antennas?

A. There are devotees to both polarizations. Over the considerable distances of shortwave signal propagation, signals have a tendency to mix and blend in polarization regardless of the polarization of the transmitting antenna that you are trying to receive; thus, the polarization is unimportant.

More important are your real estate limitations. A horizontal wire must be suspended high; even a sloper must have one end high. This requires a high mounting point or points – a tower, mast, tree, or roof peak.

When properly installed in the open, and high above the ground, a horizontal wire radiates and receives at right angles to the axis of the wire. It is virtually deaf directly off the ends. But a vertical antenna in the clear has an omnidirectional (circular) pattern for transmitting and receiving and can be secured on the ground.

Q. I've noticed that several scanner antennas, and even some HF (shortwave) antennas, are advertised as having very wide frequency ranges. Does this mean that they transmit as well as receive over these wide spans?

A. No. If an antenna is designed for reception purposes, little attention is paid to impedance matching, so if you try transmitting with it, it is likely to have high reflected power which translates to losses in the coax transmission line. Reputable manufacturers will specify the recommended, and usually reduced, transmit-frequency range as well as the receive range on their antennas.

Q. I enjoy scanner monitoring of local police, fire and ambulance calls. Is the attachable whip that came with my scanner going to be all I need?

A. For the local repeaters up to several miles away, yes. In that case, it's best to put the scanner near a window on the side facing the activity. However, if you're at all serious and want to hear some distance away, especially listening to mobiles and hand-held radios, you'll need a good outdoor antenna.

While it's tempting to mount the antenna in an attic space, that's still not quite as good as being outdoors away from noisy electrical wiring, ducting, and other metallic shields and reflectors.

Q: How can I put up an outdoor shortwave antenna that's not obvious?

A: An active outdoor antenna utilizes a small element with a high-gain amplifier. They are compact and unobtrusive.

Even less expensive and virtually invisible is wire with gray insulation. Run it out a window and over to a tree. As long as you don't have nearby power lines to add electrical noise, you will be surprised at the results.

Q. I remember reading that years ago shortwave listeners hooked their radios to bedsprings and even through a capacitor to the house wiring or telephone cabling. Are these contrivances valid alternatives to an outdoor antenna?

A. No. While these make-shift antennas work to some degree, the prevalence of household electric and electronic appliances produce enormous quantities of radio interference, especially at shortwave frequencies.

Additionally, metallic obstructions like heat and air ducting, reinforcement rods and screening in walls, metalized Mylar insulation, and aluminum siding reflect and block radio signals. You will hear strong shortwave stations, but amidst a mire of noise.

Q. Is the attachable whip that came with my scanner going to be all I need to hear local police, fire, and ambulance calls?

A. For the local repeaters up to several miles away, yes. It's best to put the scanner near a window on the side facing the activity. But if you want to hear some distance away, especially listening to mobiles and hand-held radios, you'll need a good outdoor antenna.

Q. I've heard long-time hams talk about "ladder line;" what is it?

A. It's like the old-style, flat, TV ribbon wire, only bigger and thicker! It has most of its insulation between the two parallel wire conductors punched out to reduce resistive losses, giving it a "ladder" look.

Q. There are many brands of antennas out there that have been around for quite a while. Which ones are best?

A. Actually, if the antennas have been marketed for several years, they probably have a good track record. Decide which antenna will do what you need, then buy it from an established dealer.

Antennas for Shortwave

Before we get mired in the details, let's take a general look at antenna considerations for shortwave listening, the popular, global, high frequency (HF) portion of the radio spectrum. We'll start with the most common choice, the random-length wire antenna.

First of all, because we are going to be tuning through at least a 10:1 change in frequency and thus wavelength, there's no critical, resonant length. That's why such a receiving antenna is known as a "random wire;" it isn't calculated for a particular frequency.

Second, since we're going to be receiving microvolt signal levels, any gauge wire will work just fine, so long as it's strong enough to support itself. Stranded wire is more commonly recommended because it flexes better than solid wire in the wind.

Third, a 150 foot wire is no better at receiving than a 50 foot wire. This is because the limiting factor at those frequencies is the combined natural noise from global thunderstorms – static. A longer antenna merely increases the levels of both signals and background noise, so there's no net gain in clarity of reception which is expressed as signal above noise and interference.

Fourth, the higher you can place an antenna, the better. Ground reflections, which make the response favor overhead rather than the horizon, are minimized the higher the wire is erected. And you want that horizontal distance.

Fifth, string it away from your home and power lines. This reduces the level of electrical noise interference from modern appliances and electrical arcing on old power line insulators. Avoid suspending the wire parallel to nearby electrical wiring including power lines.

Sixth, use well-shielded coaxial cable for the transmission line to your receiver. The shielding prevents pickup of electrical noise surrounding it into your home.

And seventh, erect the wire so that its broadside faces the desired target country or countries. Signals aren't being received off the ends of the wire. You can easily determine this direction by stretching a piece of string between your location and the target on a world globe. Don't use a flat paper map – the earth isn't flat no matter what you may have heard!

The dipole

In the annals of antenna history, no antenna is mentioned more than the horizontal dipole, or "flattop," and with good reason. It's easy to design, cheap to make and works very well. The term "dipole" itself simply means it has two parts, the left and the right side of its feed point (transmission line attachment).

The radiation and reception pattern would be in the shape of a fat donut with the wire antenna through the hole at right angles to the donut; in other words, at right angles to the axis of the wire. Transmission and reception in the direction of the ends of the antenna is, for all practical purposes, zero.

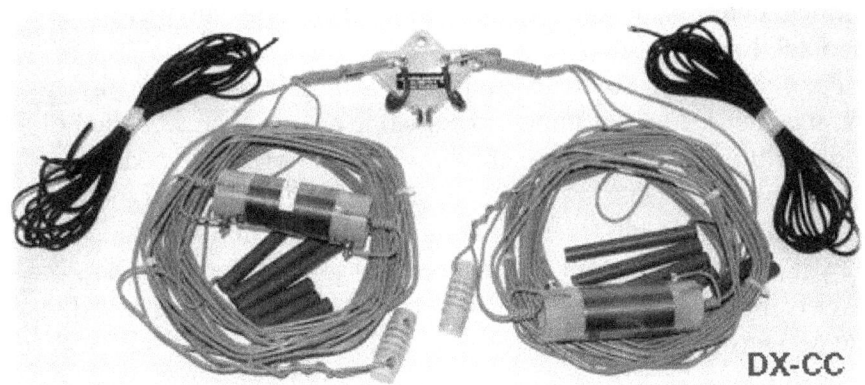

Wire antennas come in a variety of designs.

Dipoles are usually fed at the center for one good reason: If it's a half-wavelength long at the intended frequency, it matches the impedance of typical coaxial cable – approximately 50 ohms. Even 70 ohm outdoor TV coax like RG-6/U works well, and it's 100% shielded for less environmental electrical noise intrusion.

So who cares? With a transmatch ("antenna tuner"), we can make a nice, smooth, 1:1 impedance match with a dipole of virtually any length, right? Wrong.

The "conjugal match" between the transceiver and the antenna system merely provides a 50 ohm non-reactive load for the radio, but the antenna itself is still just as mismatched to the coax as it was. Reflected power will still produce standing waves (high voltage points) which waste power heating the insulation in the cable.

Voltage standing wave ratios (VSWR, often shortened to SWR), of up to 3:1 are usually tolerable with the use of low-loss coax like RG-8/U foam, RG-213, or Belden 9913, but the mere presence of high VSWR can trigger a transceiver automatic limiting circuitry to lower its output power.

There are more variables than simply the length of the antenna which determine the final impedance. Height above ground is a prominent determinant of feed point impedance for a half-wave horizontal dipole.

For example, an 80 meter dipole (3.5-4.0 MHz) has a convenient 50 ohm center feed point impedance so long as it is at an elevation of roughly 100 feet, but at appreciably lower elevations its impedance starts to rise toward a 2:1 VSWR, still quite usable.

The math is simple for calculating the fundamental length: divide 468 by the center frequency in megahertz of greatest interest; the answer is in feet. As in the example above, 468 divided by 7.1 MHz is about 66 feet.

A half-wave dipole not only matches coax impedance well at its fundamental design frequency, but at odd harmonics (multiples) as well; thus, a 66 foot dipole used on the 40 meter band (7.0-7.3 MHz) repeats that impedance on its third harmonic, the 15 meter band (21.0-21.45 MHz).

Another trick is to construct a "fan" dipole of multiple elements that all interconnect at the feed point, but gradually separate out at the ends. If the elements are all the same length, that helps maintain a steady impedance throughout a given band. If the elements are of considerably different lengths, then the antenna becomes a multi-band dipole.

But just because a multiband dipole is properly impedance matched and at the right height doesn't guarantee its directivity. The longer a dipole is, in terms of wavelength at a given frequency, the more the original "donut" shape of the emitted wave breaks up into multiple lobes. And those lobes tend to radiate more toward the ends of the wire rather than at right angles to the wire's axis.

The close proximity of reflective objects (buildings, trees, ground) can also affect feed point impedance, especially at the lower frequencies (longer wavelengths). And finally, as you tune from one end of a band to another, that also changes the feed point impedance.

A transmatch is a good investment; however, in order for it to correct the mismatch at the antenna, it would have to be physically located at the antenna feed point. This way, the 50 ohm transmission line is connected to the 50 ohm connection at the radio, providing a smooth ride to the antenna itself, and the antenna mismatch can be corrected by properly adjusting the transmatch right at the antenna feed point.

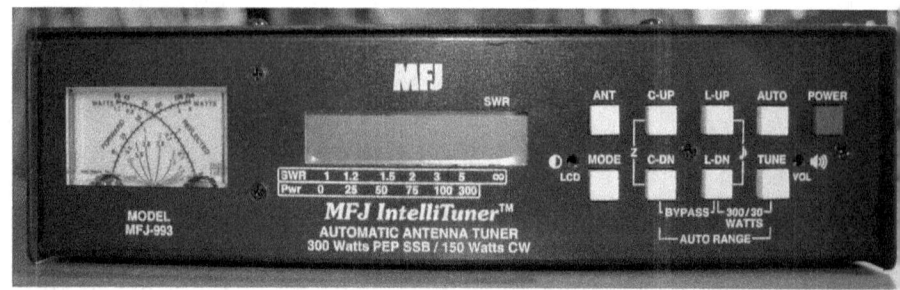

An antenna tuner matches your feed line to the radio.

Of course our theoretical antenna feed point is way above the ground, so how are we going to tune the transmatch? There are remote tuners available commercially, and some are autotuned, requiring no manual adjustments. But they are expensive.

A common alternative is to allow the mismatch at the antenna feed point, but use the lowest-loss transmission line you can afford to minimize the power loss from high VSWR. You would still need the transmatch next to the rig to minimize the VSWR there. If you have a transceiver with a built-in autotuner, so much the better.

The simplest way to deal with antenna impedance mismatch is to allow it. While this may seem counter-intuitive, it must be remembered that signal loss is due to power wasted heating the insulation of the transmission line. If we can reduce that loss, we can radiate more of our power.

Select a convenient dipole length; the choice isn't critical. Many hams select the 65 foot 40 meter half wave, but shorter or longer is just fine. Near its center attach a transmission line consisting of an open pair of parallel wires widely separated by porcelain, glass, or resin insulators. The loss is miniscule, and the match at the radio can be controlled by a transmatch.

Even the reflected power coming back down the line is eventually radiated by the antenna. It's a virtually lossless system, preferred by many hams for portable multiband operation during emergencies and Field Day, and it makes a dandy permanent antenna as well.

Another compact possibility is the trap dipole, typically a shortened 40 meter dipole interrupted on each side by parallel coil and capacitor in line with the wire element. The combination of wires, capacitors, and inductors allows several resonant frequencies, usually 80, 40, 20, 15, and 10 meters.

Multiband Verticals

Any radiating antenna element may be considered multiband if it is tuned by a transmatch, regardless of its horizontal or vertical polarization. Since we've already talked about horizontal multiband operation, let's take a look at single, vertical, multiband radiators.

If we replace one half of a dipole with a ground plane, we have a monopole. This is the basis for most vertical radiators including mobile whips. A problem with HF monopoles is that their large size dictates ground mounting, and electrically resistive earth equates with lossy signal radiation.

However, trap dipoles are shorter than full-size dipoless. They can be vertically suspended from a convenient tree limb, allowing easy installation and multiband operation in portable, emergency, and contest deployments.

Active Antennas

Because of the generally larger (or at least longer) antennas designed for shortwave, shortening the antenna is often desirable. But how can you do that and still pick up enough signals?

It was determined by a U.S. Coast Guard experiment back in the 1960s that a five-foot whip antenna provided enough signal strength to hear all stations that were in their HF network. And that was without any amplification other than the receiver's own RF stages.

With one singular exception, all antennas are passive; that is, they have no amplifying electronic circuitry. They simply reflect, refract, radiate or conduct the electromagnetic energy which reaches them.

The exception is the active (voltage probe or E-field) antenna which consists of a short (a few inches to a few feet) receiving element coupled to a wideband, small-signal amplifier. It is not used for transmitting.

While active antennas may have small size and wide bandwidth, and can deliver large signals to the receiver, they have their disadvantages. They are expensive, they require power, they may burn out from nearby lightning or degrade in performance or strong signals.

They generate noise and intermodulation interference ("intermod"), and they are usually placed close to interference-generating electronic appliances. Don't use an active antenna if an adequate passive antenna is available.

An active antenna is equipped with amplification, often 20-30 dB or so, between its junction and the coax feedline. The power source is often fed up to the antenna through the coax cable since an elevated antenna is not a good place to have to turn something on and off.

On the plus side, active antennas have wide bandwidth, high signal output, and they are compact. It would seem that active antennas have more going against them than for them, but the truth is, most of those currently on the market work very satisfactorily for SWLs who can't put up an obtrusive wire antenna.

Loop Antennas

Besides the amplified active variety, compact receiving antennas for shortwave also come in large and small open loops and ferrite rod loops which may be mounted indoors or outdoors, depending upon their construction. But the most common are small indoor models.

An active loop is equipped with a broadband preamp.

Folding a wire or a copper or aluminum pipe into a circle or square reduces the dimensional length that a wire antenna would take up. Does this smaller aperture mean poorer reception or transmission? Not by a long shot. Another benefit is its ability to be rotated to favor various directions.

By tuning the impedance with a suitable transmatch for transmitting, such a loop can be used over a wide frequency range. Even without the tuner, reception will be good over the same range.

Small loops are often used indoors for shortwave and longwave reception. For the high frequencies, the architecture is always an open loop, but at lower frequencies – AM broadcast and below – ferrite rods are frequently called into play with fine "Litz" wire coiled around them.

Once we discount the obvious disadvantage of an indoor loop being so close to electrical interference sources, they do offer a number of advantages over a simple room-strung wire, or even the short whip on a multiband portable.

Loops can be positioned and easily swiveled to minimize electrical interference and/or favor reception of specific signals. Most of them can be tuned to specific frequencies, thus reducing strong-signal overload from off-frequency signals.

When shopping for a loop antenna, pay attention to the intended frequency coverage. There are loop antennas on the market with only AM broadcast band applications (530-1700 kHz), and others that extend much higher in frequency.

HF Beams

At the higher frequencies – VHF and UHF – rotatable beam antennas are quite practical. Their shorter wavelengths allow for shorter elements and closer spacing. But at HF, such proportioning becomes unwieldy. A three-element Yagi on 80 meters would take up some 18,000 square feet of space, and how are you going to hold it up? Don't even think of adding parasitic elements.

Sure, the military has some of these monsters for intercontinental base-to-base communications, but the average radio hobbyist would have a bit of a problem mechanically, financially, and neighborly.

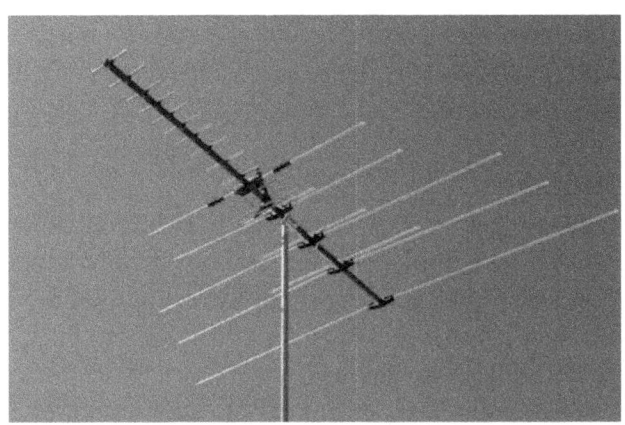

But the upper end of HF – 20 meters and higher – can be handled. In fact, it's quite common for multiband arrays to interlace several different elements on one boom, allowing automatic selection as the multiple antenna receives power at different wavelengths.

Rotatable beam antennas are readily available for frequencies above 14 MHz (20 meters). Such antennas require a husky tower and rotator, and these aren't cheap. You can gang more than one on a tower, however, so the total frequency range can be considerable, often extended well into the VHF/UHF range.

On the high frequency (HF "shortwave") bands, beams are almost always horizontally polarized. This is because distant (DX) signals are distorted into mixed polarization so that a matching antenna is not necessary.

Beam antenna elements are invariably made of aluminum tubing; it's lightweight and conductive enough to handle RF power efficiently.

The singular exception to this challenge is the wire beam, an array of three (or more) parallel wire elements following the design elements for a driven element, a reflector, and one or more directors.

Even then, such an array would have to be on high masts to avoid ground reflection which elevates the pattern, thus shortening the path toward the target area.

Vertical antennas

So far we have only discussed horizontally polarized antennas. Verticals have a well-earned place as well. They can be mounted right at ground level and many require only one support at the base. But don't mount one right against the house or you'll invite electric noise and signal reflections and absorptions.

While it's possible to simply take a dipole and mount it vertically, this requires additional suspension, and the center-fed transmission line should lead straight outward for a considerable distance to avoid distorting the pattern. This is awkward.

Far more practical is to substitute a "counterpoise" for the lower half of the antenna, allowing the single mount support and the coax to lead out along or even below the ground.

A counterpoise is nothing more than a conductive mass of metal or wires substituting for the missing lower part of the dipole. Often called a "ground plane," it is frequently made of four equally spaced, quarter-wavelength wires lying on or slightly below the ground and connected to the coax shield at the base of the antenna.

The vertical element is often a quarter wavelength at the fundamental frequency. To make it shorter, one or more loading coils may be used to add inductance, lowering its resonant frequency.

This same construction is commonly seen for VHF and UHF applications on rooftops with four drooping radial elements made of aluminum tubing or rod protruding from its base.

But getting back to HF verticals, why can't we simply attach the coax shield to a ground pipe at the base of the antenna? After all, don't we refer to "grounding" in radio?

The fact is that soil doesn't make a good, conductive ground plane. Sand is the worst. The ideal ground plane consists of 120 wire radials, each 0.4 wavelength long. But few of us can afford the real estate or the patience for that large of a ground plane installation.

Almost as effective – within 3 dB – is a field of 16 0.1 wavelength radials. The wire gauge is not critical; as thin as #18 or #20 is perfectly satisfactory if no one is likely to drive over it.

While elevated, ground-plane antennas are quite practical at VHF and more so at UHF, their quarter-wavelength elements would be cumbersome at the lower HF frequencies. But there is a way around this.

A short element may appear electrically long enough if it is connected in series with a loading coil. The inductive reactance of the coil cancels the capacitive reactance of the shortened antenna, appearing electrically as a quarter-wavelength element. Such a length reduction is useful for both elevated and ground-mounted verticals

Low-band Antennas

First, let's look at what I call the "longwire conundrum." Many hobbyists simply assume that a longwire antenna is an arbitrary, oblique reference to a wire antenna that stretches a good length. In fact, even a wire a few inches in length can be a longwire antenna if the frequency is high enough.

The "long" reference is not random, but specific in terms of the wavelength at which it is operating. If the length is greater than a full electrical wavelength at its operating frequency, then it's a longwire. Thus, a 136 foot wire is a longwire on 10 meters, but a short wire on 80 meters!

The pattern of a longwire (or any antenna for that matter) is identical for transmitting and receiving. The lower the frequency for any given wire antenna, the more its pattern favors a right angle to the axis of the wire. The higher the frequency, the more the main lobes begin to migrate toward the ends of the wire.

The pattern doesn't effectively change depending on how it's fed; center fed, off-center fed, or end fed, it's the same.

The presence of earth below a horizontal antenna causes signal reflections which alter that pattern. The lower the frequency, the higher the antenna must be erected to avoid those pattern-distorting reflections. Too close to the ground and the main lobes extend above the wire (great for working overhead airplanes) not off to the sides where we want them to reach into the horizon. Ground reflections become worse the closer the antenna comes under one-half wavelength in elevation.

Any horizontal wire at a fixed elevation above ground behaves better the higher the frequency at which it is operating, since the electrical wavelength becomes shorter and shorter.

The half-wave dipole in free space is the standard of comparison for antenna gain figures. If a multi-element antenna is specified to provide 6 dBd gain, that means it will transmit and receive a signal that is one S-unit stronger than would be delivered or received from the sides of a half-wave dipole.

Wire Antenna Directivity

While at VHF and UHF frequencies we can easily add elements to improve directivity, at lower frequencies (longer wavelengths), it's a bit awkward. The same rules apply, but the required real estate generally forbids such an array of wires. A case in point is the Wullenweber array, affectionately called the "elephant cage" for its huge dimensions. This giant contraption is obsolescent, but has had quite a military history listening for enemy communications.

A single wire can be made quite directional as evidenced by the Beverage antenna. Working especially well in the lower shortwave bands, this horizontal receiving antenna is at least one wavelength long, mounted typically ten feet above the ground, and terminated with a 470 ohm carbon resistor to ground. Its favored direction is off the far (terminated) end.

Of course such an antenna is directional, and if you want to shift its direction, you're out of luck; moving a wire hundreds of feet long on a whim is not too practical. So how about an array of such antennas if you have the room?

Directional "low band" antennas aren't razor sharp. You could put up one for each cardinal point of the compass. These four coaxial lines can be led into your radio room and connected to a four-way antenna switch. Now you're in business.

Some hobbyists do the same thing with two dipoles at right angles to one another. Since horizontal dipoles are bidirectional, you can select the best performance with a two-way switch. Two and four pole antenna switches are readily available from *MT* advertisers.

Taking Aim

Plotting the pattern of a dipole antenna is done on a globe, not on a flat, dime store map. Using a piece of string, place one end on the globe where you live. Stretch the other end to the country(ies) or continent of greatest interest. That will reveal the true bearing(s) for your antenna.

Remembering that the pattern of a dipole is equal off both sides of the axis, you will simultaneously favor the desired direction and the 180-degree opposite direction.

If you have adequate real estate, you may wish to erect two dipoles at right angles to one another, each with a separate feed line, and switch between them for 360 degree coverage. Several manufacturers offer manual switches which work well for this task. These are available for maximum amateur power and are suitable up to 30 MHz and some up to several hundred megahertz for VHF/UHF communications.

Of course the operator still needs two separate coax lines running down to the interior switch. A single antenna relay could be placed where the two dipoles cross, allowing the use of a single coax line.

A separate pair of smaller wires can be taped to the coax line and soldered to the relay solenoid for remote switching.

But how can we configure a dipole so it is unidirectional, thus concentrating its energy in one compass direction? Well, we can't. But we can make it *favor* one direction – somewhat.

Sloping

A modicum of directivity, typically 3-6 dB, can be obtained at the lower frequencies by dipping one end of the dipole toward the ground. This is usually about a 45-60 degree angle from being horizontal.

Equally important are the distance from a metal support mast and the height above the ground of the lower end.

While a sloper does have gain in a preferred direction when compared to its other directions, that gain is not as high as it would have had if left as a bi-directional, horizontal antenna.

Attic Antennas

While always a bad idea, indoor antennas are sometimes the only choice, especially in deed-restricted neighborhoods. If you must put an antenna indoors, put it in the attic away from electrical wiring and adjacent air ducts.

Few of us have the luxury of a 134 foot home in which to install a full length, 80 meter, half-wave dipole. But wire antennas can be bent dramatically to conform to their environmental limitations. They can be shaped to line the perimeter of the four walls.

As a general rule, route the wire so it never turns back on itself more than 90 degrees (a right angle bend). While not as predictable as a straight length, the results are frequently satisfactory.

Because of the unpredictability of their final feed point impedance, such configurations are most satisfactorily fed by twin-lead rather than coax to a transmatch. And if you're not particularly concerned with directivity, such an installation lends itself to multiband operation.

Nearby wiring does pose a problem, not just because it can radiate electrical appliance noise into the adjacent receiving antenna, depending upon their lengths, they absorb and reflect radio signals. Copper and iron pipes, sheet metal ducting, and aluminum siding can do the same thing.

The inverted V

The inverted V antenna is a good example of how to keep the high-current feed point away from absorptive and reflective earth by elevating it to the apex of the antenna. The ends of the drooping elements (high-voltage points) come to within a few feet of the ground where their capacitive interaction with the soil may cause some length detuning of the antenna, but little signal loss.

Don't confuse a ground-mounted, counterpoised vertical with an elevated ground-plane antenna. On the ground, we are trying to prevent radiation from being absorbed by the soil; an elevated ground-plane antenna, however, behaves more like a dipole in free space, with the radials supplying half of the antenna and forming the pattern.

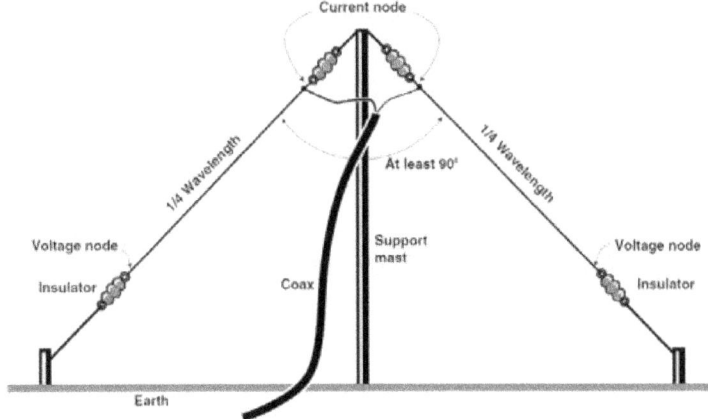

The inverted V is a popular dipole configuration.

Buried wire antennas

It is probably intuitively obvious to most hobbyists that if the ground is a loss factor in signal propagation, then burying an antenna might not be a good idea. In fact, while definitely not a good idea for transmitting due to signal absorption by moisture and minerals, reception can be quite good below a few megahertz. Electrical noise interference is frequently less as well since most of it is at nearby, elevated locations.

From ELF up through 3-4 MHz, a 100-200 foot wire buried 3-4 inches deep and end-fed by coax cable makes a dandy receiving antenna, and it's certainly not conspicuous.

Random-wire antennas

Can simply throwing a length of wire over a tree branch and attaching it to a receiver improve reception? This is one of those questions that can't be answered without asking more questions. What antenna are you using now? What kind of receiver do you have? What frequency range is of primary interest?

If you are listening in a basement apartment with an indoor antenna on a $49 portable, nearly anything will help. The rapid fluctuation of signal quality you experience when you move the portable around is a good indicator of the capriciousness of indoor antenna reception.

A 25 foot length of wire suspended from a tree branch, attached at the end to the center conductor of a length of coax will work as well as nearly any other shortwave antenna you can build or buy for general purpose listening. It makes no difference whether the wire is solid or stranded, insulated or bare, thick or thin, copper or gold, the signal capture will be the same.

But can you simply attach the center wire of that cable to the whip on a multiband portable, or even plug it into that radio's external antenna port? Shortwave portables are notorious for their poor dynamic range, the ability to treat weak and strong signals equally without overloading their signal level capacity. Overly-strong signals cause the radio's circuitry to develop "spurs," spurious signals throughout its tuning range, which aren't really there.

With a good receiver, however, any outdoor antenna will work better than a comparable indoor antenna. Outdoors, they are free and clear of signal obstructions like metal siding, metalized Mylar insulation, electrical wiring, and plumbing. They are also removed from electrical interference generated by modern household electronics.

Generally speaking, the higher the frequency being received, the less electrical interference you will experience. By the time you move up into the VHF/UHF region, electrical interference is rarely a problem. That's why scanner listeners rarely suffer from that source of interference.

But there's another reason. Nearly all electrical interference you hear buzzing from your speaker is amplitude modulated, and except for aircraft monitoring, scanner signals are all frequency modulated (FM) which is relatively immune to the raucous sound.

If strong-signal overload is a problem, a variable attenuator available from electronics outlets and online sources can be continuously adjusted from low to high levels to suit your application.

VHF/UHF antennas

In our initial part of this series we discussed antennas designed for transmitting and receiving in the high frequency (HF) part of the spectrum from 3 to 30 MHz. Except for Yagi beams and vertical antennas, the vast majority of those antennas are wire, especially on the lower frequency bands.

The tubular aluminum elements on Yagi beams lend themselves well to the very high and ultra high frequency ranges (VHF-UHF: 30-3000 MHz). Since appropriate element length and spacing are a function of wavelength, and these higher frequencies are shorter in wavelength, beams are in prominent use there.

Vertical elements are shorter as well, making elevated ground planes and collinear arrays (an inline series of vertical elements) popular at VHF/UHF.

Frequency allocations at these wavelengths intersperse land mobile scanner frequencies and ham bands with FM and TV broadcasting frequencies. For example, VHF-TV low channels 2-6 (54-88 MHz) are just above VHF low-band scanner frequencies (30-50 MHz) and six meter ham band (50-54 MHz). The VHF-TV high channels 7-13 (174-216 MHz) are just above VHF high-band scanner frequencies (150.8-174 MHz), the popular two-meter band (144-148 MHz), and just below the 222-225 MHz ham band. The UHF-TV channels 14-69 (470-806 MHz) are just above UHF scanner frequencies (450-470 MHz), the 420-450 MHz ham band, and just below the rapidly-emerging 800 MHz scanner band (806-960 MHz) which contains the shared 902-928 MHz ham band.

The point being made here is that there is some cross-over between antennas for the various services, and sometimes a good TV antenna is a good scanner antenna, and even works on the ham bands.

Polarization

The polarization of an antenna, whether vertical or horizontal, refers to the polarity of the electric portion of the wave, not the magnetic portion. As the voltage portion of the wave travels along the antenna element, it radiates its electric field in the same plane. If the element is vertical, then the electric field will be vertical as it leaves the antenna.

Because the VHF and UHF spectrums are used largely by land mobile licensees, and mobile whips are vertically polarized, virtually all VHF/UHF antennas – base and mobile – are designed for vertical application and installation.

Broadcasting, however, is a different consideration. TV antennas are horizontally polarized. But the addition of the FM band to automobile radios after WWII dictated vertical polarization for automotive whips.

This dilemma for the broadcasters – needing both vertical and horizontal polarization – led to the development of circular-polarized transmitting antennas whose signals could be received in either plane.

But over great distance, with terrestrial and ionospheric reflections and contortions, the polarization of radio waves becomes mixed. This is why shortwave signals are often received equally well with vertical and horizontal antennas.

Similar effects distort the polarization of VHF/UHF land mobile signals, especially in mountainous regions and in metropolitan areas with all the high-rise building surfaces. The orientation of the receiving-antenna element(s) becomes more a question of which position hears signals the strongest, and often a weird angle works the best!

That characteristic is easily recalled from the days of TV rabbit ears antennas which would be rotated and adjusted in angle for best reception. If you have a weather radio, you may have noticed that, depending upon its location, the desired signal may be loudest with the antenna angled from the vertical.

Directivity

All antennas share one characteristic: They are either omnidirectional (non-directional) or directional. That is to say, their field of radiation and/or reception is either uniform in all directions, or favors one (or more) direction.

Omnidirectional antennas don't have to be rotated, since their reception is uniform in any compass direction; that's an obvious operational advantage. But directivity has two advantages over non-directional antennas: They have gain in their preferred direction, and they have nulls in other directions which reduce co-channel interference.

While it may seem feasible to manually rotate an antenna, this becomes decreasingly attractive during rainstorms and cold weather! Antenna rotators are the obvious answer.

Lightweight beams are easily turned by inexpensive TV antenna rotators, but the control boxes for such devices don't have the directional resolution desired by most hams and intent scanner listeners.

Several prominent manufacturers offer heavy-duty antenna rotators for communications antennas. They are expensive, but they are long lasting, very dependable, and offer high resolution of direction.

Those Little Whips

Most, if not all, portable transceivers and scanners come with plug-in or screw-in antennas. Even the best of these suffers the indignity of not having a resonant (suitable-length) ground plane or counterpoise to complement the antenna's feed point. The radio's chassis is the closest it gets to being a complete antenna system.

Discones

The discone is often thought of as a cluster of dipole elements at an angle. It may look that way, but discones are actually waveguides which focus the emitted or received signal toward and away from the coax feed point. And they were named for their original composition: a sheet-metal disc and cone.

Discones can be designed to cover at least an 8:1 transmit frequency range by maintaining a steady feed-point impedance. This means that multiband transmitting from, say, 120-960 MHz is achievable, and even wider reception frequency range which isn't quite so impedance-matching critical.

An exponent of Second World War aeronautical communications, discone antennas have become endemic in scanner applications, and they work well for multiband ham radio applications as well. They do have two drawbacks, however. First, the higher in frequency on which a given discone is operated, the higher the takeoff angle above the horizon; while this is good for ground-to-air communications, it does limit distance for terrestrial use. And second, they have no gain'

Nonetheless, the VHF and UHF discone remains high in popularity due to their enormous bandwidth. 100-1000 MHz is a common specification, extended even further at the bottom end by an additional vertical element to add amateur six-meter (50-54 MHz) coverage. This element simultaneously provides reasonable 30-50 MHz low-band VHF land mobile reception for scanner applications.

Mobile Antennas

In an automotive environment, the vehicle body is the ground plane; its contours and antenna placement influence the radiation (and reception) pattern of the signal. Directivity generally favors the mass of metal—a roof-mounted whip has a basically omnidirectional pattern, while a rear-bumper mount favors the forward direction of the vehicle.

At frequencies above 10 MHz or so, the surface area of the vehicle and the length requirements for a vertical antenna are practical for efficient operation. A quarter-wave whip exhibits a feed-point impedance of about 36 ohms, a good match for conventional 50-ohm cable.

But as frequencies lower, the electrically-short antenna possesses less and less radiation resistance (less than one ohm at 2 MHz) and more and more capacitive reactance which must be cancelled by a series inductance (loading coil).

 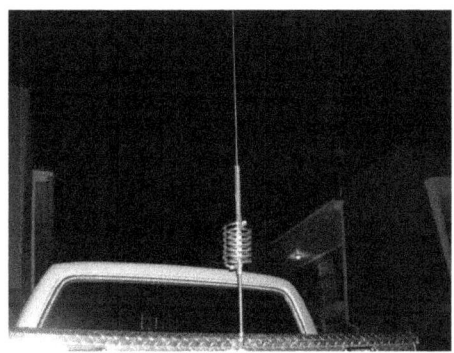

Base-loading the whip requires less inductance than center- or top-loading because the upper whip section's capacitance with the car body reduces its own capacitive reactance; but the radiation resistance remains low (in some cases a few ohms), so that coil and transmission line resistances contribute a proportionately-higher loss.

Raising the position of the loading coil changes the current distribution along the antenna, increasing the radiation resistance, but the longer loading coil requirement introduces more resistive loss. A position approximately 2/3 the way up the whip may be an optimum compromise.

Scanner Whip Antennas

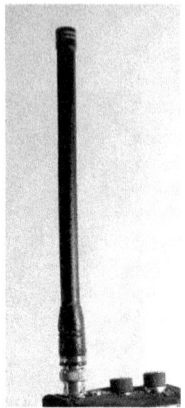

Hand-held scanners are extremely sensitive, and their specifications closely matching those of their base/mobile counterparts. So why is it that some scanners seem much more sensitive than others to weak signal response? It's probably the factory-supplied antenna.

In order to validate the claim that some aftermarket antennas will provide better performance than the original whips accompanying the scanners, we decided to put some of these antennas to the test.

For comparison, we included a conventional rubber ducky sent with a hand-held scanner. Since the least expensive antenna is a straight piece of wire, the common 18" whip was used as the baseline of comparison.

Early experiments showed that a 48" whip also does a good job, especially if 30-50 MHz low band is important.

A second test to corroborate these findings was then performed with the antennas successively tested on top of a vehicle, more reliable since it's a real-life application.

What the Test Showed

Using a simple, inexpensive, 18" whip antenna as the baseline reference, we looked for antennas that are at least equal to or better than the performance in frequency ranges of greatest interest to a scanner listener.

The 48" telescoping whip is hot at the lower frequencies when fully extended, but drops 5 dB at the 800 MHz range. Of course, one supreme advantage of a telescoping whip is that it can be adjusted to the appropriate length for any frequency in its quarter-wavelength range.

Rubber duckies from different manufacturers vary in their performance, with the longer models generally more sensitive to signals.

The simple 18" whip was outperformed only on the 460 MHz range by the rubber antennas, and low band by the 48" whip.

Multiband Dipole Clusters

Borrowing from the early TV antenna industry, some scanner antennas consist of multiple elements, some of which are parasitic, reflecting and directing signals toward the active (connected to the coax) element(s). This combination is very compact and lightweight, and provides gain over a simple vertical dipole.

Ground Plane Verticals

The ground plane antenna may be thought of as a vertical dipole in which the bottom element is replaced by an array of horizontal or drooping elements, or even a sheet of metal, simulating a perfectly conductive earth.

The vertically-polarized ground plane antenna lends itself well to VHF/UHF design. While at the lower HF wavelengths the elements can be a bit unwieldy, requiring external support, at VHF/UHF the tubular aluminum elements are shorter and lighter in weight, making self support easy.

Using thicker elements like tubing helps widen the bandwidth of an antenna, but thinner, stiff-wire elements are commonly used as well – the thinner the element, the less wind-load resistance. This is a consideration for stormy-weather-prone areas.

Multiband ground planes are also commonly available for both transmitting and receiving. They utilize the decoupling effects of coils along the vertical element to automatically choose which element is resonant for the frequency of choice.

In some cases a coil is present at the base of the vertical element to act as an impedance transformer, allowing a longer 5/8 wavelength antenna to add a couple dB of gain.

A ground plane designed for a specific frequency will experience an increase in feed-point impedance and radiation angle toward its ends (in this case upward and downward) as the applied frequency increases. This reduces its effectiveness for harmonic operation unless you are planning to work overhead aircraft and satellites.

Beams for VHF/UHF

The serious VHF/UHF signal hunter/DXer is likely to home in eventually on a beam antenna because such an antenna provides improvement in weak and distant signal reception as well as the reduction in interference of signals from off-center directions.

Hams commonly use vertically-polarized Yagi beams from 50 MHz on up. Gains on the order of 6-12 or more dBd can be achieved with additional directors (forward elements).

So, how does your transmitted signal improve with a better antenna compared with simply boosting power with a linear amplifier? For your signal to be heard one S-unit stronger, that's a boost of 6 dB (decibels) – the equivalent of quadrupling the power.

Thus, you can radiate one S-unit higher signal strength by switching from a half-wave dipole antenna to a four-element beam.

The Log Periodic Dipole Array

The LPDA derives its name from the fact that its dipole element lengths and spacing are calculated logarithmically. It is a beam in every sense including gain, and its superiority over a simple Yagi comes from its enormous bandwidth, often 10:1 or more.

Log periodics are found from the high end of the HF spectrum (notably 14-30 MHz arrays) clear up through VHF/UHF bands (1 GHz and above). They are easy to spot with their truncated triangle shape with straight edges. The best known to the monitoring hobby is the Create CPL-5130, available in two frequency ranges, 50-1300 MHz and 105-1300 MHz. It also works well as a transmitting antenna.

Those Really Big Antennas

In its simplest terms, a radio antenna is any device which is designed to emit or capture electromagnetic energy. So why isn't a screwdriver just as effective as a Yagi beam? Is bigger always better?

There are many considerations to be made in designing an antenna. At the higher frequencies above, say, 50 or 100 MHz, tricks can be played with various combinations of elements that would be impractical at the lower frequencies.

It's all a matter of the wavelength, and the lower the frequency, the longer the wavelength; thus, the longer and wider spaced the elements.

Why do world broadcasters use such large arrays other than just to handle the RF power? The answer is directivity. In order to concentrate the signal in a particular direction, elements must be phased (spaced with regard to wavelength) to enhance radiation upward (takeoff angle) to take advantage of skip propagation, and outward horizontally for extended ground-wave coverage.

The Array

When elements are all electrically interconnected, they are called "driven." If they are isolated and reflective, as on common beam antennas, they are said to be "parasitic." As an example, VHF and UHF beam antennas like the Yagi have isolated, reflective elements. Only one pair is connected to the feedline. A slightly-longer reflector is behind the driven element, and several parasitic directors are in front.

Depending upon where they are fed by the feedline, driven arrays may be either end-fire (main lobes radiate from the far end of the antenna) or broadside (main lobes radiate from the sides of the antenna).

Since the elements of driven arrays are directly fed by the transmission line, their patterns can be controlled by phasing the lines. By feeding one set of elements with a particular length of transmission line and the other set by a different length, one can change the phases between the elements to intentionally narrow and thus provide gain over that of a single, broad lobe.

One of the simplest forms of this combination of wire elements is known descriptively as a "Lazy H" – referring to its appearance as the letter H lying on its side – or even more imaginatively as a "butterfly dipole."

More accurately known as a four-element broadside array, approximately 6 dB gain can be achieved with spacing of 3/8 to 3/4 wavelength of the parallel elements.

As with all of these arrays, feed point impedance is high (hundreds of ohms), and they are balanced systems. Direct coax feed is not an option.

When two or more elements are in an axial line with each other, it's called a *collinear* array. The resulting radiation pattern is broadside (at right angles) to the elements. The number of collinear elements that can be used is theoretically endless, and some real giants have emerged for shortwave broadcasting.

Receiving vs. Transmitting Antennas

For receive-only purposes, antennas large enough to bring received signals up and over background noise (static) are usually adequate, but for transmitting, elements need to be thicker, and directivity must be taken into consideration as well. On the shortwave bands, this means a lot of metal spread over a lot of space, and the more power and sharper the directivity, the more metal and the more space.

Because of mechanical and electrical symmetry of these arrays, the impedance is the same at either end; therefore you can connect quite a series of them to get desired radiation pattern of the colossus.

Complications come not from the basic principles of the elements, but from their relative placement in order to form the desired directivity and gain. Similarly, size is not due to the basic design of the antenna, but because of the multiple identical units working in concert to achieve directional gain over a simple dipole or vertical.

Legends abound around these high-powered broadcasting behemoths, such as the homeowners near Voice of America (VOA) facilities complaining that they couldn't sleep at night because their fluorescent lights continued to glow after they were switched off!

Enter the Sterba Curtain

Collinear arrays may be daisy-chained into a large complex of antenna units. Simply stated, the antenna consists of virtually any number of collinear pairs of parallel half-wave elements spaced a half-wavelength. The end horizontal elements are a quarter-wavelength, joined by a half-wavelength element. It is a broadside array.

The Sterba has a rather narrow bandwidth and is usable only on one band of frequencies.

Variant curtains like the distributed or branch feed systems, similar to the Sterba, are popular among international broadcasters. The Sterba has a very practical application in cold climates. Because it is a closed loop of wire, an AC current can be simultaneously applied to the entire system, warming the wire to prevent it from icing up! RF chokes are used to isolate the AC current from the RF current.

The Rhombic Antenna

As a wire antenna becomes longer and longer in terms of wavelength, the major lobes migrate toward the ends of the wire, resulting in considerable gain and very narrow patterns. Phasing two or more elements together can magnify these characteristics.

The rhombic antenna gets its name from its rhombus (diamond) shape, an arrangement of end fed, multiple half-wavelength wires parallel to the surface of the Earth. It may be terminated at the far end by a resistor or it may be left open. That far end is also the direction of the major radiated lobe, so this array is an end-fire.

Since the rhombic is a balanced antenna, it should be fed by twin-lead, preferably open-wire feeders or ladder line for high power. It has high feed-point impedance, so a tuner (transmatch) is required.

If the far end of the rhombic antenna is left open, high forward gains over a simple half-wave dipole are achievable. The maximum recommended size is six wavelengths per leg, providing up to 20 dB gain – more than three S-units.

A terminating resistor is often used with a shorter rhombic (two wavelengths per leg, for example) to reduce the back lobe, thus improving front-to-back ratio. The resistance is the same as the nominal impedance of the antenna, typically 800 ohms.

The terminating resistor must be non-inductive; otherwise the coiled turns of the resistance wire would add reactance which would complicate the impedance of the system.

Gains of only 3.5 dB less than that of an unterminated rhombic of the same size can be realized with the resistively-terminated antenna. That loss is from the 1/3 of the RF power that is wasted heating the resistor!

The longer the rhombus, in terms of wavelengths, the more gain it has and the sharper the lobe. Since this large an array can't be rotated, an overly-sharp beam width is impractical. A maximum of six wavelengths per leg is recommended.

A rhombic for the 14-29.7 MHz range should be mounted at about 70 feet elevation and tilted upward to take advantage of propagation.

The bad news is revealed by the math. We are talking about a large antenna –typically six wavelengths per leg – to achieve that sort of gain. On ten meters, that approaches 400 feet from tip to tip, and just under 200 feet wide. Do you have a nearby football field that's not being used?

That's just for a "small" rhombic. A six-wavelength rhombic antenna on 160 meters would be well over a mile long and a half-mile wide.

Why so big?

Building scaled-down versions of some of these monsters for use in ham radio, and especially in shortwave listening without concern for transmitting, calls for some logic.

First of all, these antennas are all quite frequency dependent, some with quite narrow bandwidths before impedances and patterns change dramatically. Because we are talking about the shortwave band, multiple-wavelengths are considerable, resulting in substantial sizes.

In the majority of cases, gain over a simple dipole is just a few decibels. This may sound like a huge improvement, but doesn't necessarily equate to improved ability to receive. To hear the difference, simply tune in an international broadcaster and note the signal level. Now tune in another station about 6 dB stronger or weaker. Do you hear much difference? Probably not.

But if there's no immense improvement in received signal, why do the international broadcasters resort to such mammoth installations?

Huge, government-sponsored and religious broadcasting stations can afford the custom hardware required to construct these large arrays, but their primary goal is to propagate their messages to specific locations, and that's where directivity comes in to concentrate their radiated power in a given direction.

*   *   *

Hopefully, the pages of this book have taken away some of the mystery from the subject of antennas. Hobbyists often overlook the importance of a properly designed and installed antenna system. The success of transmitting or receiving, regardless of the quality of the radio equipment, is most dependent upon the antenna system.